Quick & Basic Hydronic Controls

Practical-Is-Good (P.I.G.) Technical Training Series

A Contractor's Easy Guide to Hydronic Controls, Wiring, and Wiring Diagrams

Quick & Basic Hydronic Controls

Carol Fey

P.I.G. Press
Littleton, Colorado

Disclaimer: This manual is intended as a tool for a classroom setting. Do not attempt any of the instructions in this book without the guidance of a qualified instructor.

Published by P.I.G. Press
759 E. Phillips Drive S.
Littleton, CO 80122-2873

Publisher's Cataloging-in-Publication Data
Fey, Carol.
Quick and basic hydronic controls: a contractor's easy guide to hydronic controls, wiring & wiring diagrams / Carol Fey. — Littleton, CO: P.I.G. Press, 2000.
p. cm. (practical-is-good (p.i.g.) technical training series)
ISBN 0-9672564-1-0
1. Hydronics. 2. Hot-water heating. 3. Dwellings—Heating and ventilation. I. Title. II. Series.
TH7467 .F49 2000 00-104341
697/.4 dc21 CIP

PROJECT COORDINATION BY JENKINS GROUP, INC.

03 02 ◆ 5 4 3 2

Printed in the United States of America

Dedicated to
the hearty souls
roaming the
wetheads' wonderland

Contents

Acknowledgments

Thank you to Honeywell and tekmar for technical support and drawings.

About the Author

Carol Fey is a degreed technical trainer who has worked in the controls industry for over 20 years. She has been honored as National Technical Trainer of the Year by the American Society for Training and Development. She is also the author of *Quick & Basic Electricity: A Contractor's Easy Guide to HVAC Circuits, Controls, and Wiring Diagrams*.

How To Use This Book

This book is for you if:

You're new to hydronic controls

You've tried to work with hydronic controls before and just didn't get it

You ought to know what you're doing, and don't want to ask for help

You fall asleep easily when in class or reading

When you have to read a book, it better be a short one

This book is intentionally small. If you like, you can carry it with you and read it in small pieces. For many, that's the best way to learn. Or you can take it all in at once. It's short enough that it could be done in one sitting.

There's lots of white space. Go ahead and write in it with comments and questions. It's your book.

There are lots of pictures. You're encouraged to write on the pictures, too.

The circuits and wiring presented here are examples intended for learning. You are cautioned not to actually do any wiring or work with electricity without the guidance of a qualified instructor or supervisor.

Have fun!

Introduction

Like its precedessor *Quick & Basic Electricity*, this book is purposefully small. You can gobble it up all at once. Or you can read it in bits and pieces if that's how you do best.

There's much more to say about hydronic controls than just a small book's worth. So this book is limited to "water side" hydronic controls and circuits. It doesn't cover "fire side" controls such as gas valves and oil primaries. Gotta draw the line somewhere!

Hydronic controls are what we use to deliver comfort and safety. Without controls, we could deliver plenty of heat. But heat alone isn't the business we're in. Controls allow us to deliver just the right amount of heat at just the right time. When we do that, we're in the comfort and safety business.

The circuits and wiring samples presented here are examples intended for learning. You are cautioned not to do any wiring or work with electricity without the personal guidance of a qualified instructor or supervisor.

Power Supply, Switch, and Load—Just Count to Three

Controls can appear complicated. Wiring can be messy and confusing. In this book we'll make it just the opposite. Controls will become simple. And wiring will start to make sense.

In the world of controls we can talk about new versus old. We can argue over electronic versus electromechanical. We can weigh using panels over just plain old wiring. But what it all comes down to in understanding controls is that there are just three kinds: power supply, switch, and load.

What matters most isn't what a control is called, but what it does. It's a little like on a construction job—we don't need to know a guy's name, but is he the plumber or is he the electrician?

If you can count to three, you can understand a control circuit. A circuit needs at least one of each: power supply, switch, and load. And everything in a circuit is one of these three. Isn't that incredibly simple!

Let's look at the three functions one at a time.

The power supply is where the electricity comes from. Electricity as it comes from the utility is called line voltage (about 120V). Here is a symbol for a line voltage power supply.

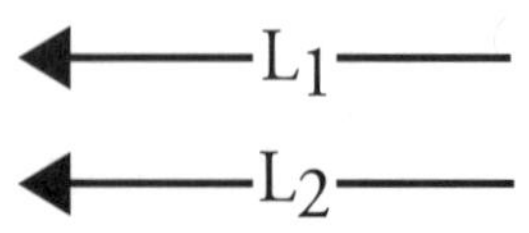

Figure 1. Symbol for line voltage power supply.

We often step down electricity from line voltage to low voltage. Low voltage is about 24V. We use a transformer to transform one voltage to another. Here's a low voltage power supply.

Figure 2. Symbol for low voltage power supply (transformer).

A switch turns something on and off. What makes a control switch different from a regular light switch is that it turns on and off depending upon a special situation. For example a thermostat turns the heat on when there is a fall in the room temperature.

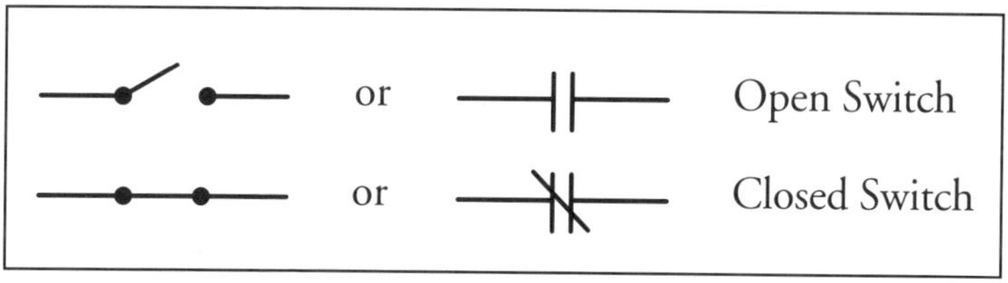

Figure 3. Switch symbols.

A load is a consumer of electricity. A load makes electricity into what we really want—light, heat, or motion. Light bulbs and motors are familiar loads.

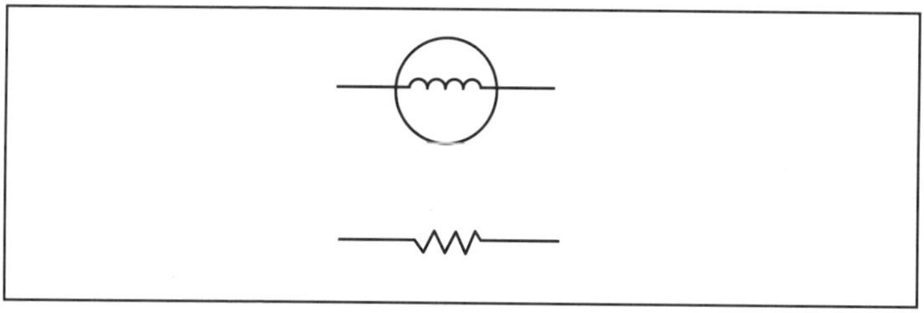

Figure 4. Load symbols.

To have a working circuit, there has to be at least one of each—power supply, switch, and load. Does that mean we can string together as many of each as we want as long as there's at least one of each? It's not quite that easy. But there are some easy rules for what you can put together.

In controls wiring, the rules may be different than if

you've taken a general basic electricity class. That's because electricity instruction is usually for electronics technicians or for engineers. We're not either of those things. We're hydronics specialists, right? Here are our rules.

There can be only one power supply in a circuit. If you have more, you may have a problem something like a cock fight.

Sometimes folks wire a couple transformers together to work as one. It's much better to use the transformers separately. That way you don't have to worry about putting them together wrong. And if one fails when they're separate, the other can keep on working.

One load's enough for each control circuit. Electronics folks like to put lots of loads (resistors) in a series circuit. Then they do math to figure out what all those loads do once they're strung together. In hydronics, it's a cold day (sometimes literally) when we put more than one load in a circuit. And this means (hooray!) you can pretty much forget about doing math.

We often put several switches in one circuit. Think of a series of switches as a safety committee. Each switch is an expert on a particular concern. There's the boiler temperature expert, known as the high limit switch, the boiler water level expert, known as the low water cut-off, and the burner flame expert, known as the flame rollout switch. Each switch votes "OK to fire the boiler" by staying closed. Then all it takes is for the "we need heat" expert, known as the thermostat, to close, and the boiler fires.

Any one of the switches has veto power. A switch exercises its veto power by opening. One open switch means the boiler doesn't fire.

Consider what a string of switches means for troubleshooting. The switches you don't know about have just as much veto power as those you find and check.

Circuits—Thinking In Circles

Circuit rule number one, as we just saw, is that there must be three things: a power supply, a switch, and a load.

Circuit rule number two is that these three things have to make a complete circle.

This means you have to think in circles (when's the last time that was a good idea!).

When thinking about controls, think about three. When thinking about circuits, think in circles. Say to yourself—power supply, switch, load, and back to power supply—complete circle. Power supply, switch, load, and back to power supply—complete circuit. It's three in a circle.

Imagine the power supply, switch and load all joining their little imaginary hands for ring-around-the-rosie. As long as they stay joined together, things can happen. But if there's a break in the circle everything stops.

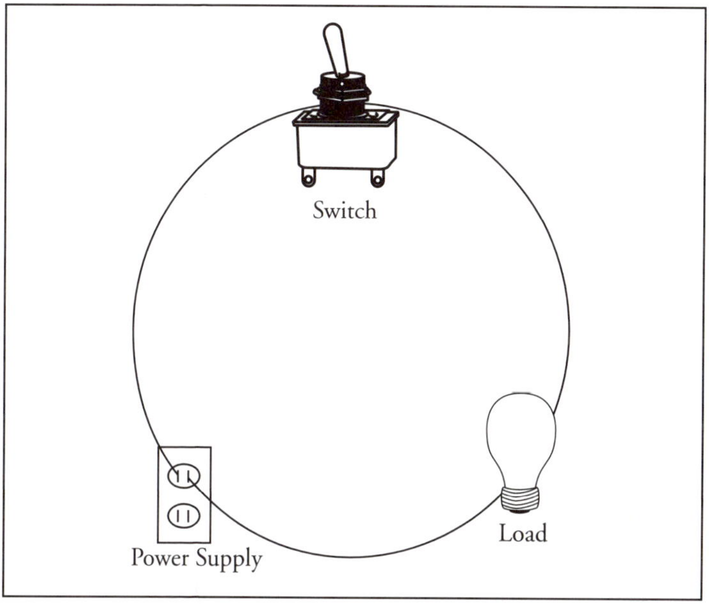

Figure 5. A circuit as a circle.

A tricky thing about circuits is that they don't look like circles. In a drawing or diagram, they look like squares or rectangles. They have corners.

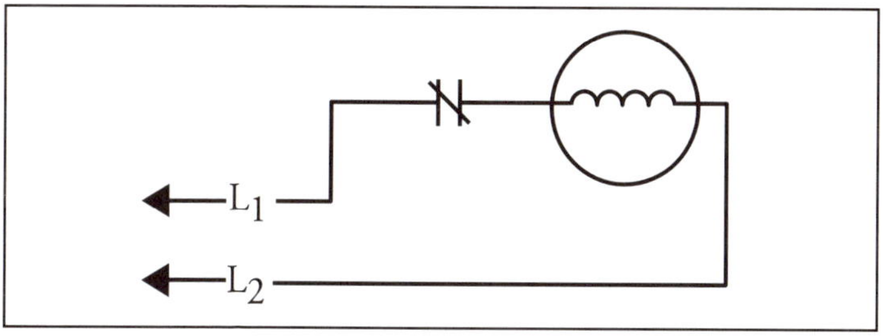

Figure 6. A circuit as a rectangle.

In real life, a circuit often looks more like a pile of spaghetti.

But you can do some easy translating. Think like a bug! Picture a rope with the ends tied together—it's continuous. Lay it on the floor. Now imagine you're a bug. If you take off along the rope in one direction, you'll eventually get back to where you started, no matter what shape the rope's in.

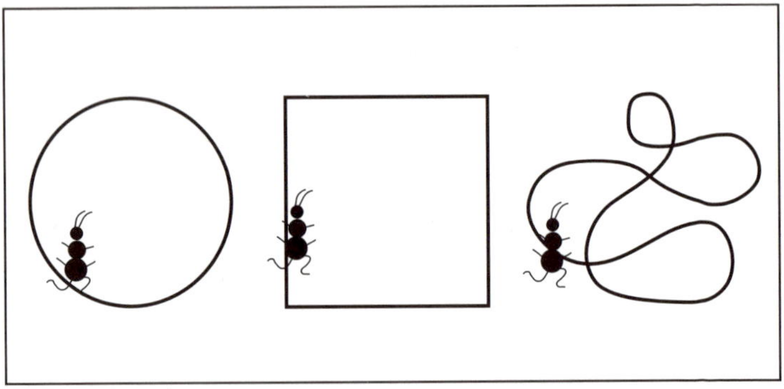

Figure 7. Test for a complete circuit by thinking like a bug on a rope.

The rope can be a circle or rectangle or even a pile of spaghetti. From the your bug's perspective it's a complete circuit as long as you can get back to where you started. It doesn't matter where or which direction you start.

If there are knots in the rope, the bug can go over them as if they didn't exist. The same goes for electrical connectors. A wire nut or other electrical connector is like a knot in a rope.

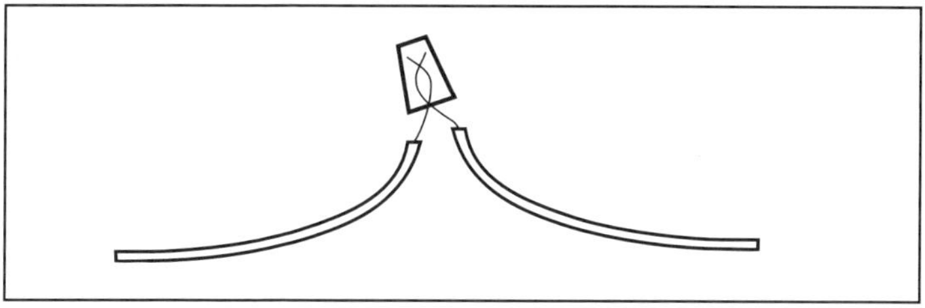

Figure 8. A wire connector.

Just as it doesn't matter to the bug if the rope is a different color in various places, it doesn't matter to electricity what color a wire is.

One big difference for electricity, though, is that the path has to be metal (or water, but that's not a good idea).

Circuits can be wired in series or in parallel. This terminology refers to how loads are put together. Let's take a quick look at what each means. For more detail, you might look at the book *Quick & Basic Electricity* by this same author.

Here is a series circuit.

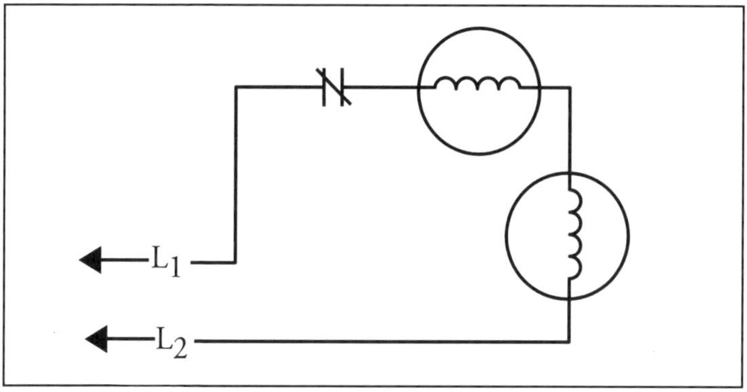

Figure 9. A series circuit.

In a series circuit loads are strung together like beads on a necklace. The classic example of loads in series is old-fashioned Christmas tree lights. If one goes out, they all go out.

A much better way to wire loads is called parallel. You can still wire several loads and their switches to one power supply. But each load has its own private access to the power supply. In a string of lights, this means that if one goes out, the others keep on working.

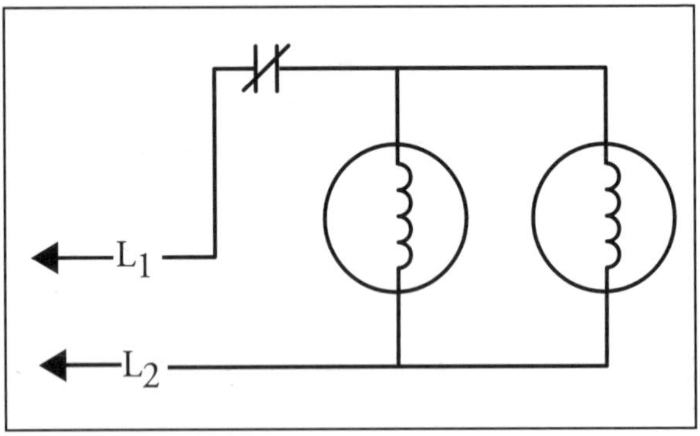

Figure 10. A parallel circuit.

These loads are in parallel with each other (notice how the loads are on parallel lines). Strictly speaking, they are all in one parallel circuit. That's because they all have the same power supply.

For our purposes here, though, we're going to call each group of power supply, switch, and load a circuit.

Power Supplies

The place where electricity comes from is called a power supply.

There are so many things we can get tangled up in when we think about power supplies. There are all those voltages. And there's the confusion about phases and number of wires (as in single phase three wire as opposed to three phase four wire, etc). Let's leave this electrician stuff to the electricians.

What we need from the electrician is line voltage somewhere around 120V. The electrician might even deliver that line voltage into a transformer that steps down the line voltage to low voltage. We call low voltage 24V, even though it's not always exactly that amount.

Controls usually have just two voltage options: line voltage and low voltage. Both are alternating current (AC). We don't use direct current (DC), which comes from batteries.

In an electrical diagram, the line voltage power supply is shown like this:

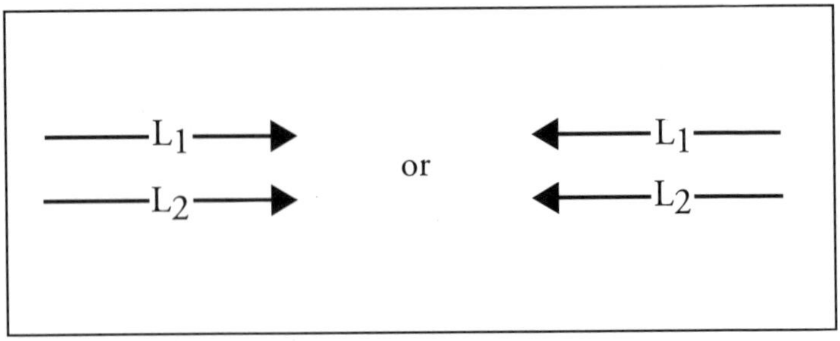

Figure 11. Symbols for line voltage power supply.

Think of L1 and L2, plus the two lines with arrows, simply as the symbol for line voltage power supply.

[Interesting but non-essential information: There's no need to over-analyze this symbol. But if you have the kind of mind that needs a symbol to make logical sense, then this explanation may be interesting. First, if the rule is that a complete circuit can't have any breaks, how come there's a break between those two arrows? For an answer you get to use your imagination. Imagine that the circuit actually goes all the way back to the power plant where the electricity originates. Check out the picture below—you'll never see it this way anywhere else, but if it works. . . .

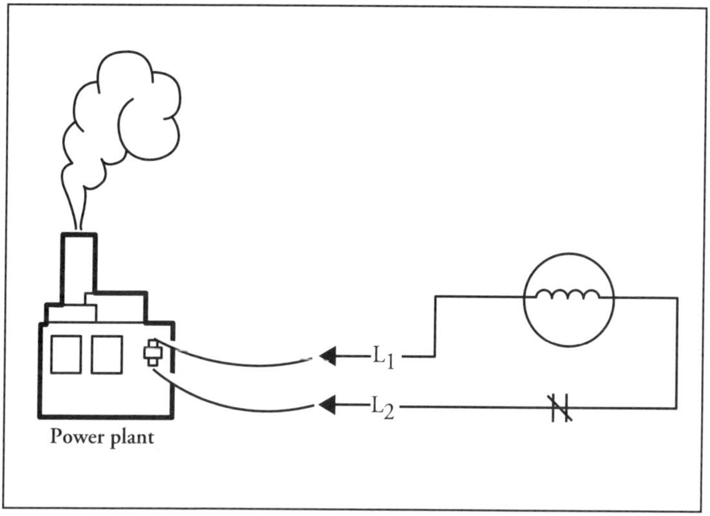

Figure 12. Line voltage circuit with power plant.

Second, we're used to thinking that arrows show the direction of movement. It's tempting to assume those arrows show the direction that electricity is moving. Give up that idea. That's not what these arrows are doing. If you really need for them to be pointing to something, imagine they're both pointing to the power plant-the source of the electricity.]

We use low voltage rather than line voltage for most for most hydronic heating controls.

[Important but non-essential information: One reason for using 24V for controls is that low voltage allows us more accuracy for temperature control. Low voltage controls let us keep room temperature within a couple degrees of temperature setpoint. It's been found that most people can't sense less than two degrees difference between the warmest that the room gets and the coolest. Line voltage controls give more of a temperature difference (called temperature swing). When room temperature varies more than two degrees, people say things such as, "My, it's getting warm in here," or "Crank up the heat, would ya."

Never forget that when dealing with hydronic heating, we're in the comfort business.]

The step down transformer we use to take 120V down to 24V looks like this:

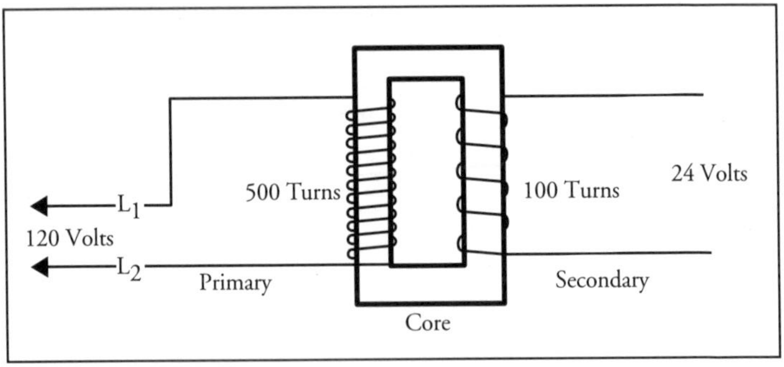

Figure 13. Step down transformer.

A transformer has two sides. The primary (just like in school, primary means first) side is usually shown on the left side. It's where the electricity first comes in. The primary is typically 120V. The voltage is going into the transformer.

The right side of the drawing shows the secondary. It shows the electricity after it's been stepped down. This side is typically 24V. This voltage is coming out of the transformer. The secondary side of the transformer is the power supply for a 24V circuit.

The chunk of iron in the middle of the transformer is called the core. Transformers are one of the few things where heavier still is better. Heavier means more iron, and more iron insures reliable transfer of voltage from the primary to the secondary.

[Interesting but non-essential information: If you're wondering how primary voltage gets transformed to secondary voltage, the simple explanation is magic. A more scientific explanation is electromagnetism.

Somewhere back in elementary school you may have had a science lesson where you learned you could wind a wire around a big nail, power the wire, and turn the nail into a magnet. The primary side of the transformer is like that wire, and the core is like that nail.

We call that wire wound around the core the primary winding.

On the secondary side of the transformer there's a completely separate winding. It's called the secondary winding. There is no physical or electrical connection between the primary and secondary windings. The only relationship between the primary and secondary is magnetic.

Look at the transformer drawing again. On the left side is the primary at 120V. On the right side is the secondary at 24V. The drawing says there are 500 turns or windings of the primary wire around the primary side of the core. And there are 100 turns on the secondary side. The relationship between 500 turns and 100 turns is 5 to 1. Or, there are 1/5 as many windings on the secondary as the primary. So we could expect there to be 1/5 as much voltage on the secondary as the primary. It works—1/5 of 120V is 24V, just what we were looking for.

For the record, transformers can also be step up.

In a wiring diagram, the low voltage power supply, or transformer, is shown like this:

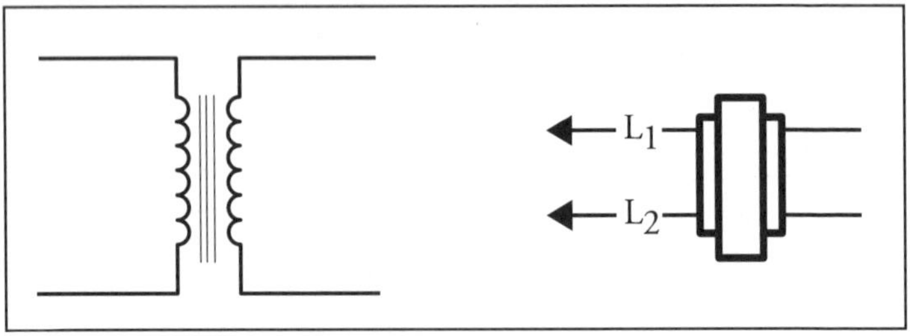

Figure 14. Transformer symbols.

A transformer's size or capacity is given as a VA rating. VA means volts multiplied by amps. Because this is a low voltage circuit, we can assume that we have 24 volts.

Each user of electricity is called a load. Each load has a known appetite for electricity, known as an amp draw. A load's amp draw is usually written on the device itself and in the technical literature. Hold that thought.

A 40 VA transformer is typical in a controls circuit. If we divide 40 VA by 24V, we know how many amps this transformer can handle. 40 VA divided by 24V equals 1.66 amps. This means that the total amps of all the loads connected to this transformer must be less than 1.66 amps.

You can overload the transformer when you try to draw too much amperage through its secondary side. That's called burning or smoking the transformer.

Let's do a little calculation to figure out how many zone valves we can put on one 40 VA transformer. 40 VA divided by 24 V, as we saw above, is 1.66 amps. Each zone valve that we're using has an amp draw of .32 amps. Divide 1.66 amps by .32 amps and we get an answer of 5.18. Round it back to 5 zone valves.

There's always an old timer with another answer. If your old timer is says less than five, well, no harm done. If you old timer has always gotten away with more than five, here's what's probably going on. Not all of the zone valves have called for heat at the same time. When they do—and that's going to be in the middle of the darkest, coldest night—that transformer's going to burn. Then none of the zone valves will work until you replace the transformer. Replace it with a bigger transformer. Or install two 40 VA transformers with no more than five .32 amp valves on each.

Different models and manufacturers of zone valves can have vastly different amp ratings. Consider the difference between an amp rating of .32 versus .8. If your valves are of the .8 amp variety, you can use only 2 of them, not 5, on a 40 VA transformer. Big difference!

Hydronic Controls that Are Switches

Many hydronic controls are simply switches. You know them as thermostats, aquastats, or limits. A clue that a control is a switch is the syllable -stat.

A switch can turn something on or off. Each stat in a hydronics system has the purpose to sense temperature, and to turn something on or off if that temperature isn't just right. If the temperature is correct, nothing happens. That correct temperature is called setpoint.

Each stat has a setpoint. The setpoint can be fixed (permanently set), or it can be adjustable. The temperature dial on a thermostat or aquastat controller allows it to be adjustable.

Hydronic controls that are switches fall into two categories: operating controls and safety controls. The function of an operating control is to maintain a desired temperature. A thermostat maintains the desired room temperature. An aquastat controller maintains boiler water temperature.

An aquastat controller functioning as a safety control is called a limit. Rather than maintaining a desired temperature, its purpose is to prevent an undesirable temperature. A high limit turns the boiler off if it goes above a certain setpoint. A low limit turns the boiler on if it goes below a certain setpoint.

A stat changes its switch position on either temperature rise or temperature fall. Think about an ordinary thermostat. The switch is normally open. Upon temperature fall below setpoint, the switch closes, or makes, to bring the heat on. Upon temperature rise above set point, the switch opens, or breaks, to turn the heat off. The switching action is make on temperature fall and break on temperature rise.

[Interesting but non-essential information. You may have heard that electricity is just like water. This "water analogy" makes sense only so far. Here's where it stops working: when a valve is open, stuff flows through it. When a valve is closed, stuff stops. But in an electrical circuit, open means there is a break in the electricity's path, as in opening. An open switch is off.]

Switching action varies depending on the function of the control. Most safety controls are designed to be normally closed. That is, if conditions are safe (normal) the switch stays closed. A high limit switch has the function to open (turn things off) if the temperature gets too high. So its switching action is to break on temperature rise.

Differential is the term for the temperature difference between where a stat makes and where it breaks. If there weren't a difference, the switch would rapidly flip between open and closed. Differential is either fixed or adjustable. A good room thermostat has a fixed differential less than two degrees. That means that there are no more than two degrees between the warmest the room gets and the coldest.

Aquastat controllers typically have a larger differential. When they come with boilers they often have a fixed differential. Replacements often have an adjustable differential so that they can be used in a variety of applications.

There are several ways for stats to sense temperature. Sensing can be either electronic or electromechanical. If the stat is electronic, sensing is done by a thermister. Its resistance changes as temperature changes. Switching takes place as resistance changes.

Electromechanical stats depend upon the principle that substances expand as they are heated, and contract as they are cooled. The element that does this is called the bimetal. A bimetal gets its name from the fact that it is two (bi- means two, as in a bicycle has two wheels) metals sandwiched together. Temperature changes cause the bimetal to move in a predictable way. In a lower quality thermostat, that movement does the switching directly.

In a higher quality electromechanical thermostat, the bimetal moves a mercury bulb, which then does the switching. This allows much more accurate temperature control. Mercury is used for two reasons. Since it's a liquid, it reacts to a very small amount of bimetal movement, or temperature change. And since mercury is a metal, it conducts electricity through the switch.

An electromechanical aquastat controller also depends upon the fact that substances expand when heated. The sensing element, or bulb, is filled with a temperature-sensitive material. The bulb is attached to the controller by a thin tube called a capillary. As the material in the bulb is heated, it expands into the capillary tube. As it expands it pushes on a diaphragm inside the aquastat. The diaphragm movement causes the aquastat switch to open or close.

An electronic aquastat controller, like electronic thermostat, senses temperature with a thermister.

Switching action can be described in terms of poles and throws. This switching action is usually identified with four letters, for example SPST or DPDT. S stands for single, D for double, P for pole, and T for throw. The terms single and double mean the same as they would anywhere else.

A pole in a switch is the connection between two contact points. Think of a fishing pole. If you hold one end and your buddy holds the other, you have a single pole. If you have two more buddies standing beside the first two of you, and a separate pole connects them, you have double pole(s). There is no electrical connection between the two poles. In switches, the two poles move at the same time.

Keep thinking about fishing (when did someone last tell you that?). Imagine it's just you this time holding a pole. If you cast, or throw, so that the pole connects you and your buddy, that's single throw. If you could also cast behind and connect with another buddy, you would have double throw. You can connect with one buddy or the other, not both.

If you have a double pole, your throws can be either single throw or double throw.

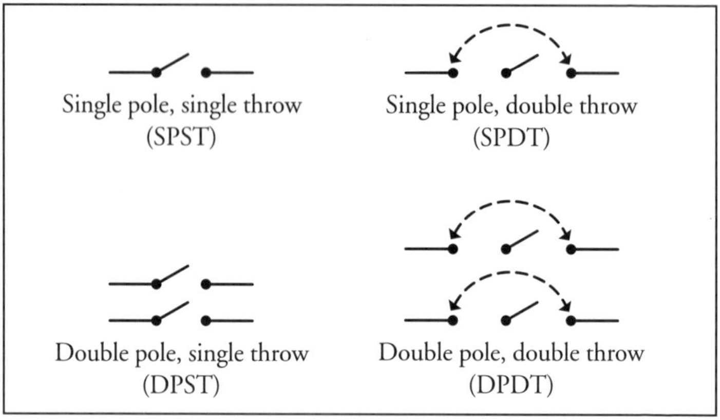

Figure 15. Switch poles and throws.

A two-wire zone valve is controlled by a thermostat with single throw switching. The thermostat, not the valve or the valve motor, is the switch. When the thermostat switch is closed, power goes to the zone valve motor, which opens the valve. When the switch opens there is no power to the zone valve motor. With no power to hold the valve open, a spring in the valve pulls it closed. This is like a screen door with a spring attached. You have to power it (push it with your hand) to open it. When you let go, the spring pulls it closed.

A three-wire zone valve is controlled by a thermostat with double throw switching. When the thermostat switch is closed in one direction, power goes to the valve motor so that the motor opens the valve. When the switch is closed in the other direction, power is applied so that the motor closes the valve. Thinking again in terms of a screen door, this time the door doesn't have a spring. You can either push and hold it open, or you can push and hold it closed.

Actually, there's no such thing as a four-wire zone valve. It's a two-wire zone valve plus two wires for an end switch. The end switch is not part of the valve, even though it's attached to it.

Using the end switch is completely optional. When the valve becomes fully open, it closes the end switch. This means that you can use the end switch to turn on anything you want—the end switch doesn't care what. Impractical examples would turning on a light bulb or ringing a door bell.

A practical use for the end switch is to turn on the boiler circulator. The circulator needs to run only if a valve is open. So it's useful for each valve to be able to say to the circulator, "I'm open, so get pumping!"

The zone valve motor and the end switch are in completely different electrical circuits. The two circuits can have different voltages. For example the zone valve motor is probably in a 24V circuit. The circulator though is probably in a line voltage circuit. That works because there is no electrical connection between the two.

Modulating switching means that a switch can open the valve a little, a lot, or somewhere in between. Terms with similar meaning are floating and proportional control. Switching devices can be called controllers rather than stats, but they're still switches. A modulating valve must be controlled by a modulating controller. That's because a conventional switch can choose only fully-open or fully-closed.

Multistage switching means bringing on two different heat sources, called stage one and stage two, at different times. If there is need for a little heat, only stage one comes on. If more heat is needed, stage two joins in. A multistage thermostat is like two thermostats in one. In an electromechanical multistage thermostat, you can actually see two separate bimetals and mercury bulbs.

Programmable means that the heating setpoint can automatically change, depending upon the time of day or night. It's sometimes called a night set-back or clock thermostat. Like a non-programmable thermostat, a programmable may be either electronic or electromechanical.

Cycle rate is the number of times per hour that the boiler can come on if it's needed. Six cycles per hour (every ten minutes) is the industry standard to assure maximum room comfort. You get six cycles per hour when you set the thermostat anticipator to match the amp draw of the load in the circuit.

Typically the only load in each hydronic zone circuit is a zone valve or zone pump. Look on the device to find the amp draw or amp rating. It will be a decimal, such as .2,

.32, or .8. Inside an electromechanical thermostat you'll find a dial with decimal numbers on it. This is the anticipator. Simply set the anticipator to match the amp rating of the valve or pump. Remember, the purpose of the anticipator is to set the cycle rate. The purpose of the cycle rate is to have even temperature. That means comfort.

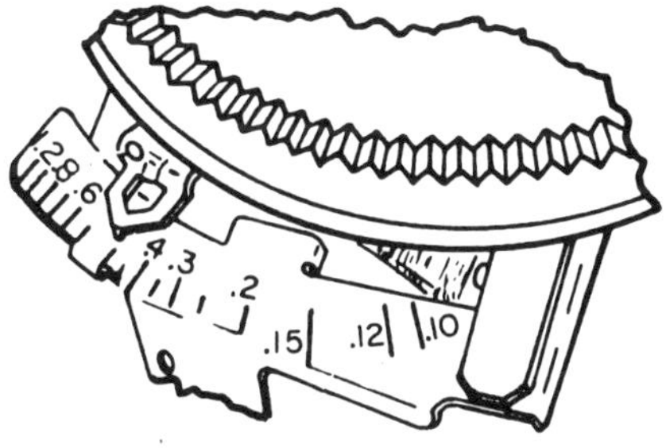

Figure 16. Match the thermostat anticipator setting to the amp draw of the load.

[Interesting but non-essential information: Most people are not aware of a temperature swing of less than two degrees. It has been found that this two-degree range is best maintained by giving the heating a chance to come on every 10 minutes (60 minutes divided by six cycles per hour = 10-minute cycles). In error some folks think that they might save money by cycling their heat less often—maybe only every 20 minutes. But look at the chart of how temperature swings increase as we decrease from 6 cph to 3 cph.]

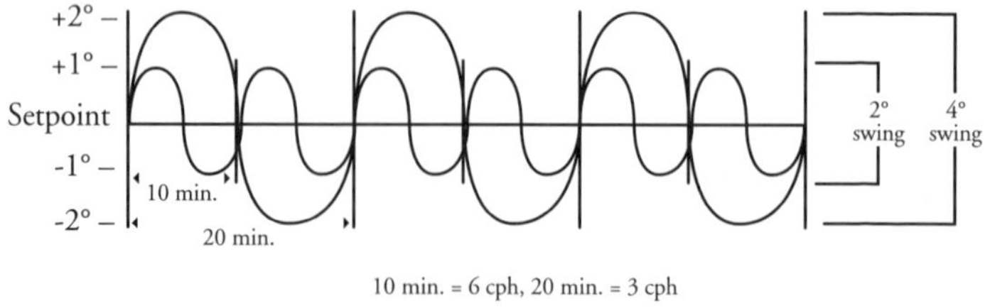

Figure 17. Cycle rates and temperature swing.

A cycle is on-time plus off-time. When there is little demand for heat, the equipment runs only a short part of that ten-minute cycle. If there is no need for heat, it won't run at all. When there is a great demand for heat, the equipment will run a greater portion of that cycle. If necessary, it could run all the way through one cycle and continue into the next.

For various reasons, we may want to use a cycle rate other than 6 cph. Equipment manufacturers specify cycle rate for new installations. For older hydronic systems with heavy cast iron radiators, 3 cph is often advised. This doesn't apply to baseboard hydronics—leave that at 6 cph.

Electronic thermostats don't have an anticipator. They have a cycle rate adjustment, which does the same thing. They are usually factory set for the standard 6 cph. Follow the accompanying instructions if you want something different.

You can change the cycle rate on conventional thermostats by adjusting the anticipator setting. For 3 cph, simply multiply the amp rating by 1.2 and set the anticipator to that number.

Cycle rate for in-floor radiant heat is an industry debate. Here's how it goes. Old-style cycle rate wisdom says that the bigger the mass of the heat conductor, the lower the cycle-rate. By itself, that makes sense. A larger mass gives off heat longer than a smaller mass. It needs to be re-heated less often. That's why we cycle fin-tube baseboard more often than a cast-iron radiator.

Continuing this thinking, we could decide that a floor is an even larger mass than a cast-iron radiator. Perhaps in-floor radiant should have an even slower cycle rate. Some in the industry went to thinking 1 cph. Others went so far as to say that a non-anticipated thermostat would be ideal. However, non-anticipated thermostat would be like controlling heat with a toggle switch.

When we look at what's really going on, we may need to do something else entirely.

What cycle rate is really about? Cycle-rate does two things. It determines how often there's an opportunity to correct the temperature in the room. And it determines how much temperature swing (difference between warmest and coolest temperature) there will be. As there are fewer opportunities to correct the temperature, there is more temperature swing. If we consider this resulting discomfort along side the overall purpose of in-floor radiant heat (comfort), we have a conflict.

Certainly, a floor is bigger than a cast iron radiator. But is it really always a bigger heating mass? In the case of staple-up radiant heat, the wood floor isn't a large heating mass. Furthermore, there may be considerable heat loss on the underside of the floor.

Another problem with in-floor heat can be over-heating, particularly on sunny days.

Back at our theoretical one cycle per hour, let's say that the heating system cycles and delivers an hour's worth of heating capacity to the floor. Then the sun arrives and adds more heat. The room over-heats.

Before we resort to an air conditioner, let's look at the cycle rate alternative. By adjusting the cycle rate, we can have more opportunities to correct the temperature. As room conditions change, the temperature can be corrected. This way of thinking points to more cycles per hour.

Even though in-floor heating was used as far back as the Romans, as we use it today it's still a rather new industry. The way is open to experiment with cycle rate as a tool to deliver maximum comfort.

A snow melt sensor is another kind of switch. It senses for two conditions simultaneously: temperature and moisture. Think of the snow melt sensor as two switches in series. When the temperature gets low enough, one switch closes (close on temperature fall). When there is snow or rain, the other switch closes. That means that both switches have to be closed before the snow melt system comes on. Snow melt can also be controlled by a manual timer.

Outdoor reset, or outdoor temperature compensation, changes heating system water temperature based upon change in the outdoor temperature. A primary purpose of reset controls is comfort. Boiler water hot enough for an extremely cold night may deliver too much heat for a sunny day. Another purpose is to conserve energy.

The older electromechanical reset control looks a lot like an aquastat controller with two bulbs. One bulb senses boiler temperature, and the other senses outdoor temperature. As the outdoor temperature rises, the boiler temperature is allowed to fall. And vice versa.

The part of outdoor reset that folks get confused about is reset ratio. But it's only a matter of deciding among a few limited choices. For example a reset ratio 1:1 means that for every degree that outdoor temperature drops, the boiler temperature increases one degree.

The newer electronic outdoor reset controls are sometimes called outdoor temperature compensators. They may look more complicated because they have added features, but they can be simpler than the older electromechanical models. The simplest newer models figure the reset ratio for you.

Hydronic Controls that Are Loads

A load is a consumer of electricity. It uses electricity to create motion, heat or light. In hydronics the load is usually the motor of a valve or pump. Or the load may be the coil of a relay. In that case, the coil uses electricity to create magnetism, which in turn causes switching action in a different circuit.

Here are some symbols for loads.

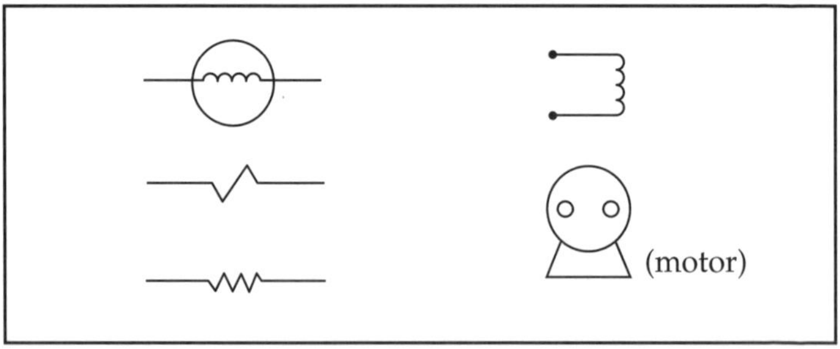

Figure 18. Symbols for loads.

You can expect a load to be shown as a figure that would be difficult for electricity to get through. You might think of it as a squiggle. This makes sense because it's the fact that the load resists electricity passing through it that causes the action. The squiggle often isn't drawn in a motor symbol—you have to imagine it. But the space is there for the squiggle regardless.

A valve functions as a door or gate. Like doors, some valves are designed to be either completely open or completely closed. These are called two-position valves. Other valves are designed to stop at various positions in between. These are called modulating or throttling.

Hydronic valves can be manually-operated or motorized. We usually adjust a manual valve, such a ball valve, to a position and leave it there. Motorized valves are automatically adjusted by a controller. When a thermostat senses a need for heat, it tells the motor to open the valve.

In a motorized valve, the water portion is completely separate from the electrical part. Let's look at the water part first.

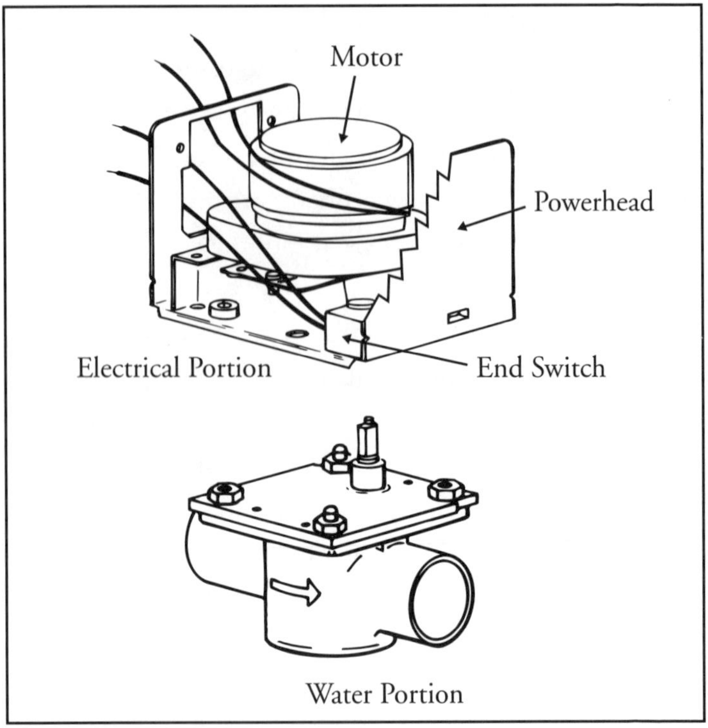

Figure 19. The water and electrical portions of a motorized valve are completely separate.

The water portion of the valve is called the body. The opening inside the valve is called the port.

The amount of water that can flow through a valve is called the flow rate or cv rating. Flow rate is measured as cv. The definition of cv is "gallons per minute (gpm) at a pressure (called pressure drop) of one pound per square inch (psi), at 60 degrees F." A typical flow rate is 3.5 cv. That means that at 1 psi, 3.5 gpm can flow through the valve.

Some folks use the term full port to refer to larger capacity zone valves. No matter how large the port or capacity, the valve can have only as much water going through as the piping will allow.

A valve is also sized by the diameter of its pipe connection, such as 1/2" or 3/4". Valves with different piping may be identical on the inside, depending upon their cv rating. Different sized pipe connected to them, though, may make a difference.

Close-off rating is a measure of a valve's ability to close against water flow. Close-off rating is given in psi. If the flow of the water is stronger than the close-off rating, the valve can't close smoothly or completely. The close-off rating depends a lot upon how the valve closes. The turning action of a cartridge is much stronger than the swinging-gate action of a ball or flapper plug.

The electrical part of the valve is called the operator or powerhead. It contains the valve motor. In zone valves up to about 1", the powerhead and the body are contained in the same unit. Larger valves may have a separate motor, and a linkage to connect the two.

Either way, no water comes in contact with the electrical portion of the valve. Notice that wires go only to the top part of the valve, and water goes to only the bottom part.

A typical zone valve is a two-way, straight-through valve. The water comes in one side and leaves the other. Nothing much happens in between. The valve serves as an open or closed door.

Mixing valves are three-way or four-way valves. Water of different temperatures comes into the valve, mixes together, and leaves the valve at a desired temperature.

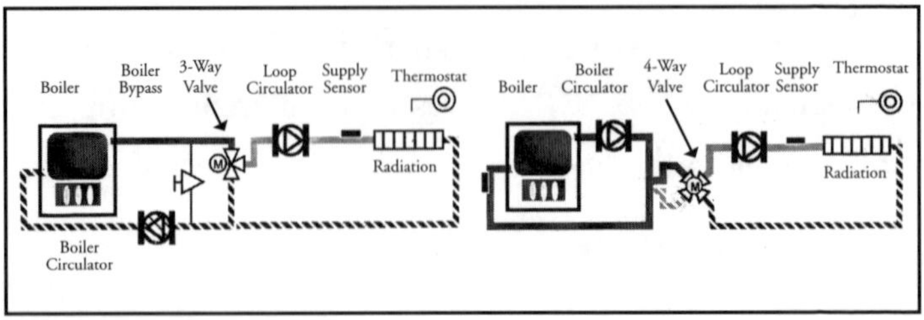

Figure 20. Three-way and four-way valves.

A diverting valve is also a 3-way valve. The water enters the valve from one direction, and leaves in two separate directions.

Zone valve noise or water hammer is usually caused by one of two situations. One is the "bigger is better" approach to pump sizing. An oversized pump is likely to cause noise when the valves try to close, especially if only one of several valves has been open.

The other common cause of water hammer is valves installed backwards. If there's an arrow on the valve, that's the way the water's supposed to go. (It may not be going the way you think it is.)

To the Boiler Room

Let's imagine ourselves in the boiler room looking at the hydronic controls. On the wall is a horizontal row of small silver boxes, called zone valves, with copper pipes coming out of them. To the left of the zone valves is a transformer. There are wires everywhere. Some connect the zone valves to the transformer. Some go from the zone valves up the ceiling. Some go from the transformer up to the ceiling. And some go from the zone valves to the boiler. Whatta pile of spaghetti!

Let's sort out one circuit at a time. A circuit is always a power supply, switch, and load. Here is a transformer, thermostat and zone valve circuit. Let's call it a single zone circuit, because it controls the heat of one zone in the living space.

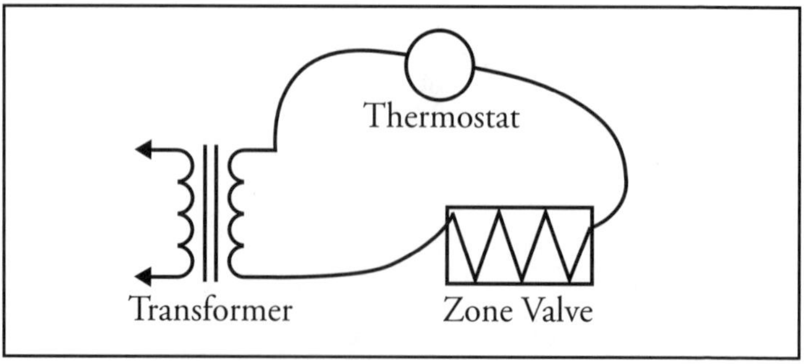

Figure 21. A single-zone circuit.

Starting at the transformer, one wire connects the transformer to one side of the zone valve. Another wire connects the other side of the zone valve to the thermostat. A third wire connects the other side of the thermostat to the other side of the transformer. We have a complete circuit.

In reality the thermostat is in the living space. That means that from the boiler room, the wires going to and from the thermostat look like they're just heading for the ceiling.

Let's add another zone.

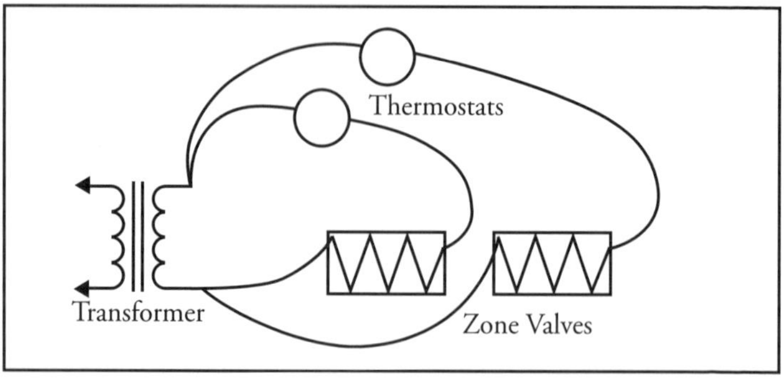

Figure 22. A two-zone circuit.

For the time being, forget about the first circuit. But let's use the same transformer. A new piece of wire connects one side of the transformer to the second zone valve. A second new piece of wire connects the second side of the second zone valve to its thermostat. A third new wire connects the second side of that thermostat back to the second side of the transformer.

Both sides of the transformer now have two wires. Each of the two wires on one side connects to a different zone valve. Each of the two wires on the other side connects to a different thermostat.

Let's add one more zone, making a three-zone system. Now there are three wires on each side of the transformer.

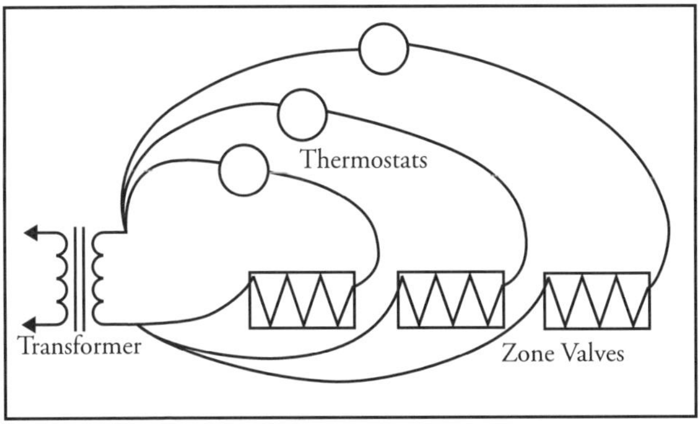

Figure 23. A three-zone circuit.

With all these wires on each side of the transformer, the transformer terminals could get mighty crowded. There are a couple of ways to fix this problem. So far we have one solid piece of wire between each of the controls. More often, though, we join wires together using some sort of connector, such as wire nuts.

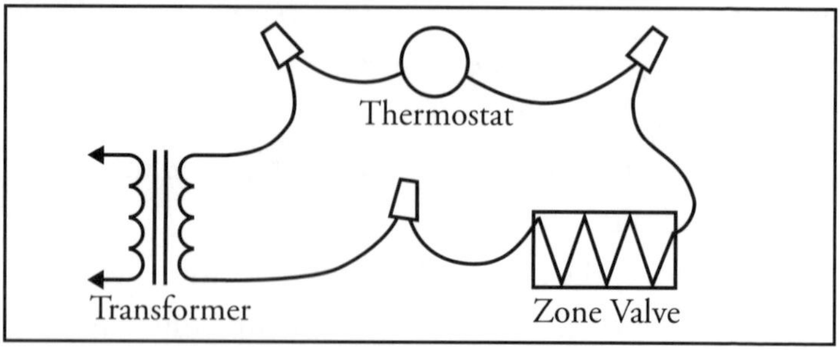

Figure 24. For electricity, a solid wire is the same as two wires connected together.

For convenience we can run a single wire from each side of the transformer. Then we can connect a wire from each valve to the wire from one side of the transformer. And we can connect a wire from each thermostat to the other transformer wire. Notice that there are four wires going to each connector.

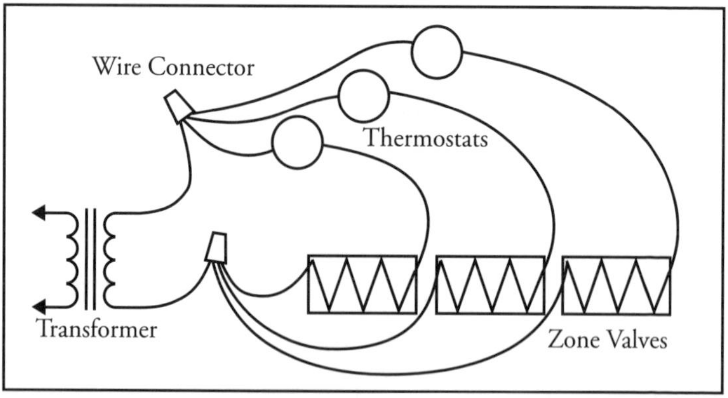

Figure 25. Circuits with wire nut connectors.

A terminal strip is another kind of connector.

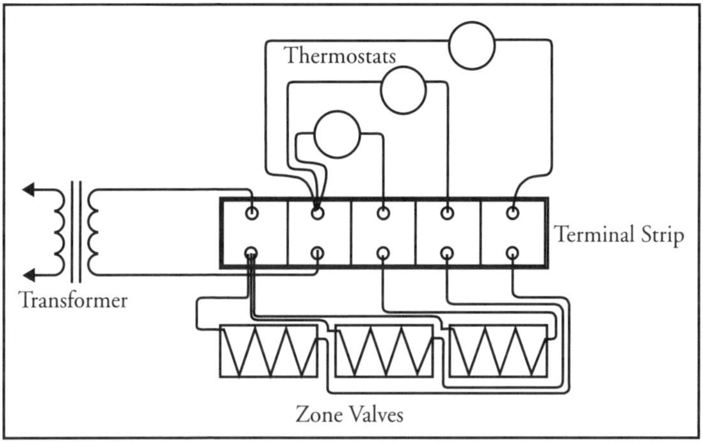

Figure 26. Circuits with a terminal strip.

The screws on a terminal strip are paired by small pieces of metal. There is no connection with any of the other screws because the pairs are isolated from each other by plastic barriers. Each pair of terminal screws can be used as a completely separate connector.

A terminal strip can create the illusion that a job is complex. But it actually makes a job easier to wire. And it simplifies troubleshooting.

A zoning panel is another way to make connections and organize wiring. Read about it in the chapter about panels.

By wiring these circuits of transformer, thermostat, and zone valve (power supply, switch, and load), here's what we've set up. When the thermostat senses a need for heat, its switch closes. The closed switch lets power flow through the zone valve motor. The motor opens the valve.

Water isn't going to move through the valve just because the valve is open. We need a circulator to force the water. But we need the circulator only when a valve is open, that is, when there's a call for heat.

Many zone valves do this with an end switch. When the valve completely opens, it closes the end switch. Notice the simple circuit made by the end switch, circulator, and line voltage power supply. You can use line or low voltage, whichever the load requires.

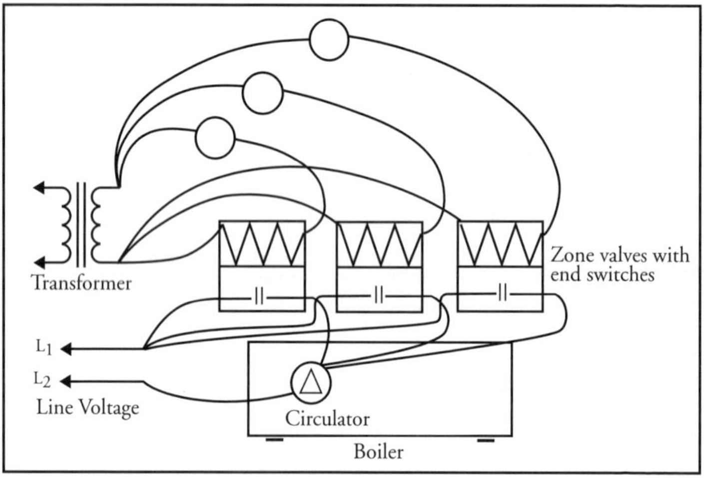

Figure 27. The end switch and circulator circuit.

Heating boiler water requires a circuit to fire the burner. That circuit is another set of power supply, switch, and load. The load is the gas valve or oil primary, the switch is likely to be an aquastat controller (refer to the chapter on switches), and the power supply is either line voltage or 24V to match the load.

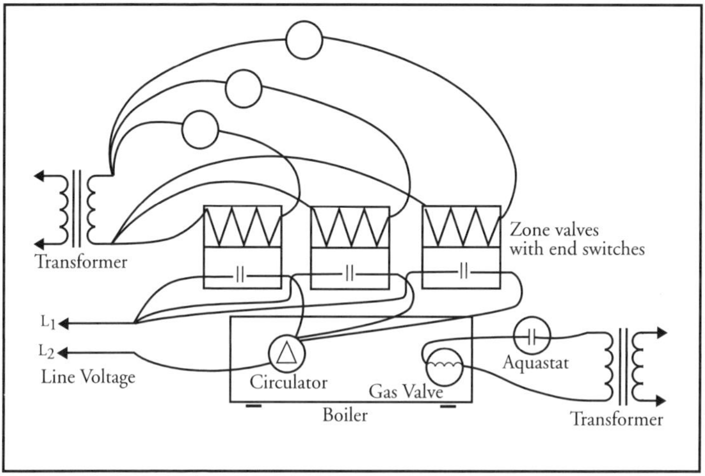

Figure 28. Burner circuit.

Relays

When a control seems especially complex, it may performing more than one function. That's true of a relay. A relay is a combination of a load and several switches.

Here's why we use relays. A circuit can have only one voltage. With that in mind, consider how we might control line voltage equipment using a low voltage thermostat. Low voltage controls provide greater comfort, but equipment is often line voltage.

Think of a relay as a translator. If I can speak only Greek and you can speak only English, we need a translator who speaks both languages. If the thermostat is only 24V and the circulator is 120V we need a relay.

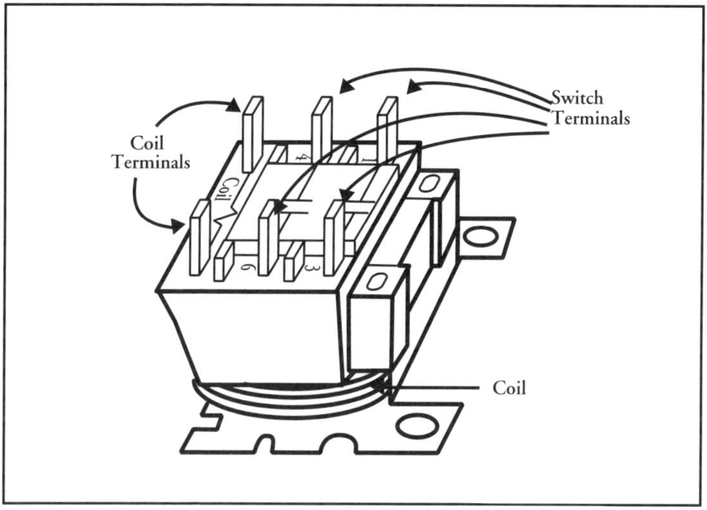

Figure 29. A relay.

In this drawing you can see the parts of a relay. The coil is at the bottom of the relay. The coil is a load because it is a winding of wire that consumes electricity. The terminals to connect to the coil are on the top of the relay (left side). The other terminals on the top of the relay are for switches. Sometimes they are called contacts. There are two terminals for each switch. In the drawing here, the terminals across from each other go together, not the ones that are side-by-side. You don't have to use all the switches.

Each switch and the coil are in separate circuits. Because there is no electrical connection between the relay parts, each circuit's voltage can be different.

Let's look at the circuits for the coil and one switch.

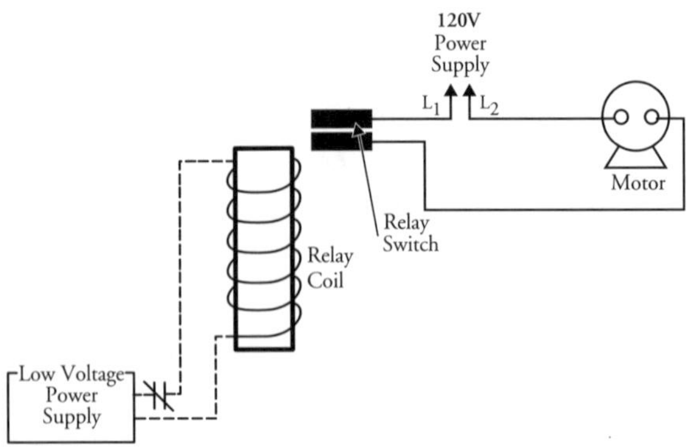

Figure 30. A relay in two different circuits.

The dotted line portion is the low voltage (24V) circuit. Look for a power supply, a switch, and a load. The transformer is a low voltage power supply. The load is the coil of the relay. Let's assume that the switch is a thermostat.

Here's what happens in the circuit. Low voltage power is available from the transformer. When the thermostat switch closes, it allows power to flow through the relay coil. We now say that the relay coil is powered. When the coil is powered, it creates electromagnetism. This magnetism causes any switches inside the relay to change position. If the relay switches are open, they close. If they are closed, they open. There can be some of each.

Now let's look at the solid line portion of the above diagram. This is a line voltage (120V) circuit. The power supply is line voltage coming from the power plant. The load is a motor. Let's say it's a circulator motor. The switch is a relay switch. The terminals for the relay switch are physically on top of the relay. The working part of the switch is inside the relay, not visible to us. The motor will run as soon as the relay switch closes.

Think backwards now. What's it take to close the relay switch? The relay switch closes when the relay coil is powered. The relay coil is in the low voltage circuit. What's it take to power the relay coil? The relay coil is powered as soon as the switch in the low voltage circuit closes. We've already said that this switch is a thermostat.

Look at the same sequence again, forward this time. The low voltage thermostat switch closes upon a call for heat. This allows power to flow to the relay coil. When the relay coil is powered it creates electromagnetism.

The magnetism causes the switch inside the relay to close. When that switch is closed, power in the line voltage circuit flows to the motor. The motor is now on, and will stay on until the thermostat switch opens. (Remember an open switch is off.)

In a wiring diagram, relay parts are labeled. The coil may be labeled relay coil. Or it may be labeled simply 1K. The switches may be labeled relay switches. Or they may be labeled 1K1, 1K2 etc. If there were additional relays in the same diagram, the second relay coil could be labeled 2K. Switches associated with coil 2K would be labeled 2K1, 2K2, etc. Note that 1K2 and 2K1 are not the same.

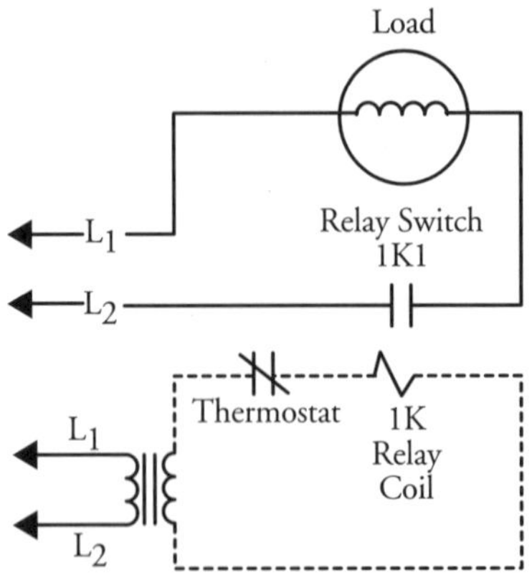

Figure 31. Labeling in a relay wiring diagram.

Relay switches are referred to as normally open (N.O.) or normally closed (N.C). Normally means when the coil is not powered. The switches (and there usually are more than one) on the top of the relay might be labeled either NO or NC. Or they may be labeled with switch symbols.

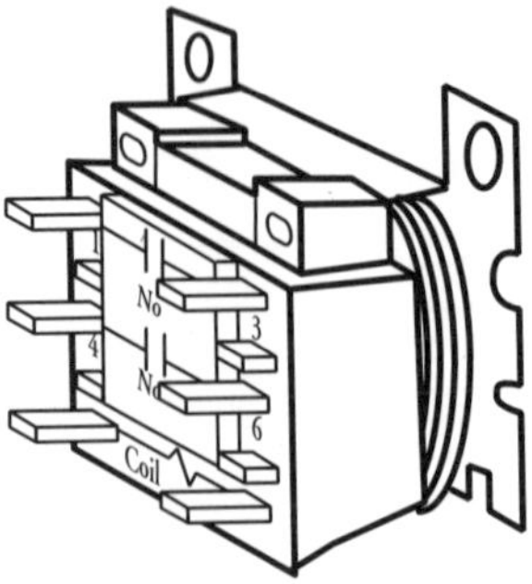

Figure 32. Relay switch labeling.

The tricky part about relays is this: as soon as the coil is powered, the switches are no longer in their normal position. When the relay coil is powered the switch position is the opposite of what the label says. That means that a normally open switch is closed. (We're not in Kansas any more, Toto!)

A relay in a wiring diagram often looks like this:

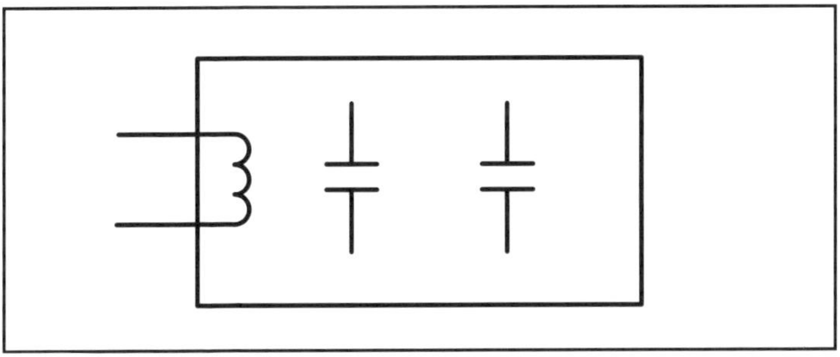

Figure 33. Relay symbol.

As we might expect, the coil looks like a load and the switches are drawn as switches. As soon as the coil is powered, both of these open switches become closed.

Panels—What's Inside this Box?

What goes on inside those mysterious boxes called panels? Until we understand them, it's easy to think that they are unnecessarily complicated. But once we do understand them, they make a job easier to install. And they organize a job so that it's easier to understand the work of someone on the job before us.

Before looking inside a box, think back to the beginning of this book. Remember "Count to three" and "Think in circles."

"Count to three" is a reminder that everything inside the panel is a power supply, a switch, or a load. In any circuit, there has to be at least one of each.

"Think in circles" is a reminder that everything works in complete circuits. Like a bug on a rope, electricity has to be able to follow a path and end up back where it started.

The first box we'll look at is an aquastat relay. It is the

combination of an aquastat controller (limit), a relay, and a transformer—all mounted together in one device.

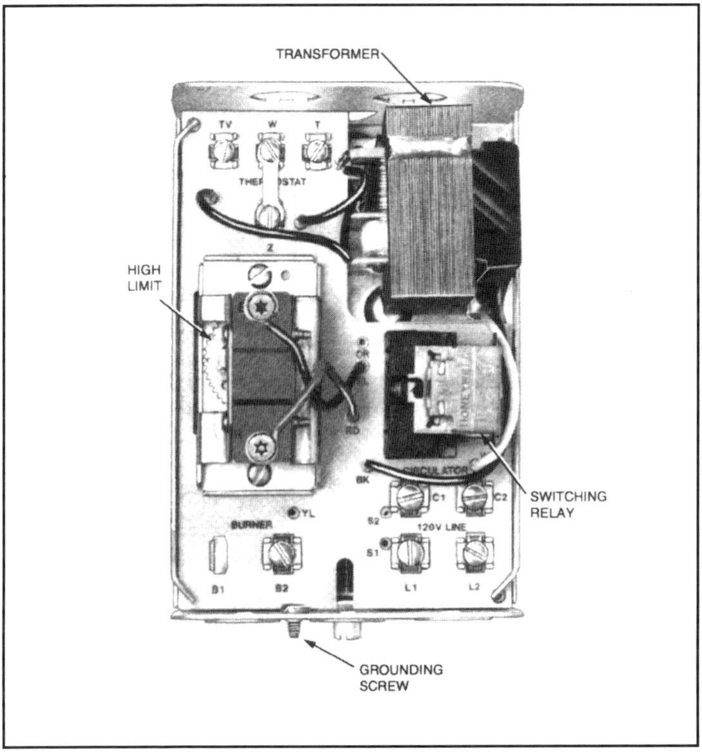

Figure 34. An aquastat relay.

To see how these individual controls work together, we can use two diagrams from the control's technical literature. One is the external connection diagram. The other is the internal schematic diagram.

Let's look at the external connection diagram.

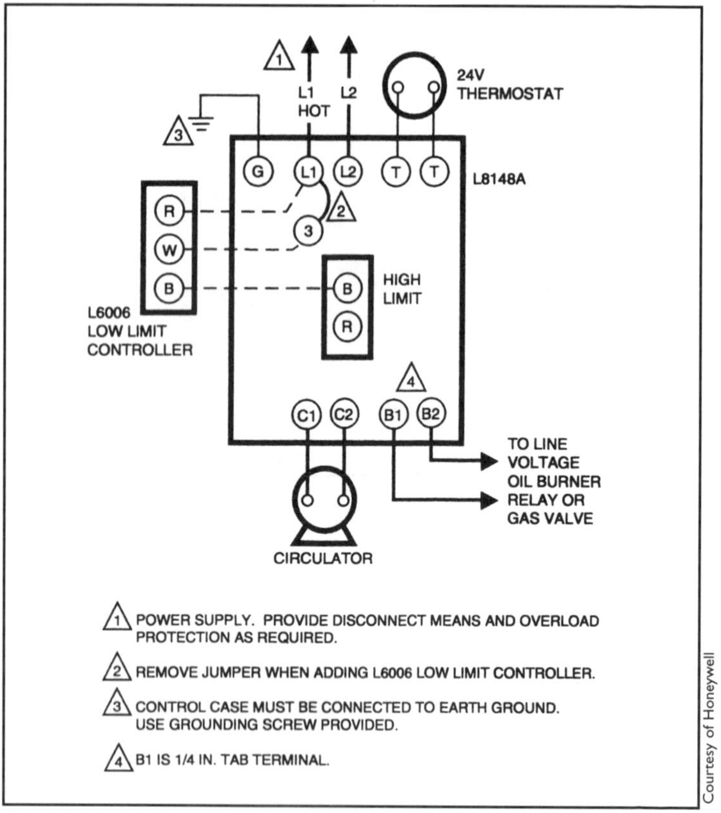

Figure 35. External connection diagram of an aquastat relay.

The big rectangle is the outline of the control. Everything inside the rectangle is visible on the control itself.

The lines going outside that rectangle show connections to other devices. They are shown as the thermostat, circulator, and oil burner relay or gas valve.

Let's start with L1-L2 at the top of the control. As you know, those two terminals go to line voltage. There's a number 1 in a triangle next to the line voltage symbol. This means there's a note for you to read. It's extremely important to read all the notes, especially if you think you don't have the time. The notes will save you time in the long run.

At the upper right corner are terminals T-T. These connect to the two thermostat terminals.

Use only the R and W thermostat terminals. On a heat-only thermostat, these are the only two you'll have. Ignore any others unless there is an "C" terminal on a digital thermostat. In that case, connect C to the 24V on the transformer.

Looking at the bottom right corner of the control, connect B1 and B2 to the line voltage oil burner relay or gas valve. If you're using a low voltage gas valve, this is the wrong control.

Moving around the control, connect C1 and C2 to the boiler circulator.

The dotted lines on the left side of the control show that adding a low limit controller is optional. Be sure to read all the notes. Note 2 in the triangle says that you must remove a jumper if you add this control.

In the upper left corner of the control, terminal G must go to earth ground.

Grounding is easy to overlook. Sometimes a job works fine without grounding. Beware, your luck probably won't last. Intermittent means it'll work when you're there, and it'll fail after you leave.

There are two kinds of grounding. Earth ground means that you must insure that there's a connection between this terminal and actual earth. It doesn't have be to a direct connection, but it has to get there eventually. The old standard is to connect to a water pipe. That works if you're sure that all the pipe going to ground is metal, not plastic. You also have to be sure that teflon tape isn't going to get in the way.

The other kind of grounding, used in electronic ignition systems, is a burner ground. In that case, you need to run a wire only from the ground terminal to the boiler burner—a short trip.

The internal schematic diagram shows what is going on inside the control. This is a picture of exactly the same control that you were looking at before with the external diagram. It's just another way of looking at the same thing.

The technical literature may show the internal schematic separately from the external connections as in the figure below. Or it may show the two in one drawing.

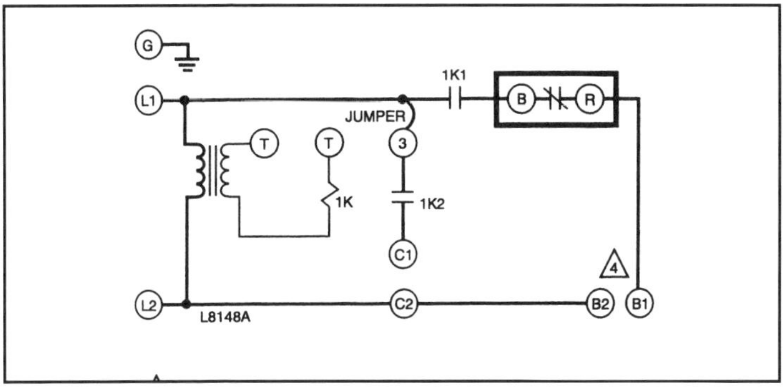

Figure 36. Internal schematic diagram.

The terminals that you see on the internal schematic are labeled the same as on the external diagram. But they may be shown in a different position.

The lines on the internal schematic show the internal circuitry of the control. This circuitry is inside the panel—you never see it.

[Important but non-essential information: Some lines in a schematic are heavier than others. Heavier lines mean line voltage, and lighter lines mean low voltage. Sometimes dotted lines are used for low voltage. Other times dotted lines mean "optional," such as in the external connection diagram above. The way for you to tell what the dotted lines represent is to study their relationship to the rest of the drawing.]

Starting with the upper left corner, there's the G terminal going to ground.

On the left side of the internal schematic are terminals L1 and L2. They always go to line voltage. L1 and L2 are the primary side of the transformer.

The other side of the transformer is the secondary. Of course, it's low voltage. The other clue that we're working with low voltage here is that this side of the transformer is shown as a lighter line.

Look for the power supply, switch, and load in the circuit that includes the transformer secondary side. The transformer secondary is always a power supply. The switch is the external thermostat wired into terminals T and T. You could pencil in that switch between terminal T and T. The load of that same circuit is the squiggle labeled 1K. The 1K label means that this load is the coil of a relay. Some drawings would label this as "relay coil."

We have a complete circuit: low voltage power supply to thermostat (switch) to relay coil (load) and back to power supply. Let's ask three questions about this circuit.

When is power available? Whenever power is supplied to the control at L1-L2.

When is the load, relay coil 1K, powered? Whenever switch T-T is closed.

When is switch T-T closed? Whenever the thermostat switch is closed, which is whenever there's a call for heat.

Let's move on to terminals C1 and C2. We know from the external connection diagram that these two terminals are for the circulator. The circulator is a load. Pencil it in the diagram if you wish. If nothing else, you know it's a load because it's drawn as a motor in the external connection diagram.

What's the circuit that powers the load called circulator? Look for a power supply. Since the lines connected to C1-C2 are heavy, we know it's in a line voltage circuit. That means that line voltage L1-L2 is the power supply. If you trace to the left from C2, you get to the primary (line voltage) side of the transformer, which is electrically the same as line voltage.

The solid dot where the two lines connect just to the right of L2 means electricity passes through as if it were a solid line. This dot might represent a wire nut connection, a terminal connection, or even a solid wire. Electricity doesn't care which.

Continue to trace upward toward L1, and through the connector. Continue to the right toward the "jumper."

Jumper means a connection that may be removed in certain circumstances. For the time being, let's leave the jumper there as a connector.

We're all the way around to the switch now. Closing the switch between terminals 3 and C1 completes the circuit. What's it going to take to close that switch?

A hint of how to close that switch is its label lK2. The 1K label means it belongs to relay 1K. So what it takes to close switch 1K2 is for relay coil 1K to be powered. What's it take to power relay coil 1K? From before, we know that between terminals T-T is a thermostat. A call for heat closes the thermostat switch, which powers relay coil 1K . When coil 1K is powered, switch 1K2 closes.

Closing switch 1K2 completes the circulator circuit. Look again. The power supply is line voltage into the panel, the switch is the relay switch 1K2, and the load is the circulator connected to terminals C1 and C2. Simply, a call for heat turns on the circulator.

There's one more load to power with this control. At terminals B1-B2 the external connection diagram says to wire in the line voltage oil burner relay or gas valve. This label or the heavy line in the internal schematic tells us that it's in a line voltage circuit.

The line voltage power supply always is L1-L2. Tracing this circuit from B2 toward L2, we come upon C2. Electricity will pass through C2 on its way between B2 and L2, so for this circuit, we can ignore C2. Trace through all the connectors up toward L1, and back along the top of the diagram to the right to switch 1K1. Again the 1K label means that this is a switch associated with relay 1K. When relay coil 1K is powered (when the thermostat calls for heat), switch 1K1 closes (as well as switch 1K2). You might pencil this switch closed. We have one more switch (B-R) to get through in order to get back to terminal B1.

The drawing shows switch B-R closed. In the external connection diagram, B-R is labeled "high limit." Switch B-R will stay closed unless the boiler gets too hot.

If the boiler does overheat, it's the high limit's job to open switch B-R If switch B-R opens, it breaks power to whatever load is across terminals B2-B1, be it a burner relay or gas valve.

[Important but non-essential information: If you're wondering about that jumper, look at what it connects. It's between terminal 3 and the line that goes to L1. That means it connects terminal 3 to L1. Go to the external connection diagram and see what's happening between L1 and 3. The dotted lines in the external connection diagram show an optional low limit controller. Remove the jumper. Connecting L1 to low limit terminal R, and terminal 3 to low limit terminal W, forces the circuit to go through the low limit instead of the direct path through the jumper.

A limit controller is a switch. So the circulator (the load at terminals C1 and C2) is powered only if the switch between R and W in the low limit is closed. If we look at the specs for the low limit, we find that R-W is closed unless the boiler water temperature falls below setpoint. This keeps the circulator from moving water that isn't hot enough.

This is the "circulator controller" portion of the controller.

Switch R-W-B is what's know as "single pole, double throw." When the connection between terminals R-W opens, the switch between R-B closes. This is the "low limit controller" portion of the control. Using the external and internal schematics together, notice that the electrical circuit would be L1 to R, to B in the low limit, to B in the high limit. As long as the high limit stays closed, then, whenever the boiler water temperature falls below setpoint, the circulator is disabled and the burner turns on.

When water reaches setpoint, the low limit controller switch opens R-B and closes R-W. This brings on the circulator if there's a call for heat. The burner and circulator stay on until there's no more call for heat. Then 1K1 shuts down the burner and 1K2 shuts down the circulator. If the boiler water gets too hot, R-B in the high limit (not the low limit) opens and shuts down the burner. The circulator continues to run until there's no more call for heat.

Now let's apply what we've learned to a different model of the same control. The main difference is that this one powers a low voltage gas valve.

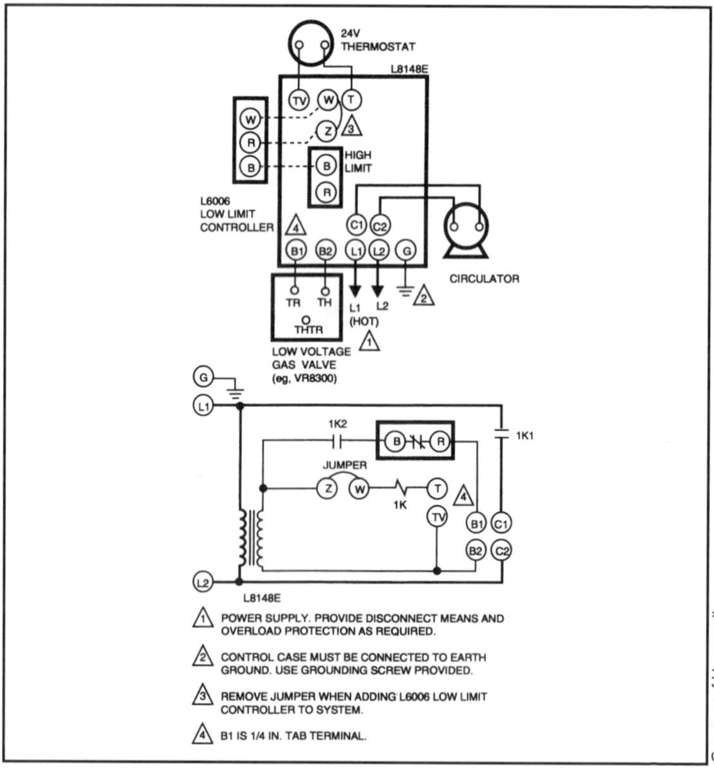

Courtesy of Honeywell

Figure 37. External and internal drawings.

First look at the external connection diagram (top drawing). Where does line voltage come in? (L1-L2 at the bottom of the diagram)

Does C1-C2 connect to a power supply, switch, or load? (load—the circulator)

What terminals does the thermostat connect to? (T and TV)

What terminals does the low voltage gas valve connect to? (B1 and B2)

Now look at the internal schematic (bottom drawing). Is the circulator in a line voltage or low voltage circuit? (line voltage)

What are the power supply and switch in the gas valve circuit? (transformer secondary is power supply and there are two switches: relay switch 1K2 and terminals B-R of the high limit.)

How does the thermostat control the gas valve circuit and the circulator circuit? Notice that these loads are two different voltages. (The thermostat is the switch in the 1K relay circuit. That relay operates switch 1K2 in the gas valve circuit and switch 1K1 in the circulator circuit.)

[Interesting but non-essential information: Chances are that there is more than one zone wired into this control panel. That means there is more than one thermostat. All the thermostats can be wired in parallel with each other so that each individually can bring on this system. Or, if zone valves with end switches are used, the system can be wired with the end switches in parallel in place of the thermostats.]

An electronic panel such as a zoning panel isn't that much different from an electromechanical panel. You can't see the internal circuitry, but you couldn't see it in the electromechanical panel, either.

When you look inside an electronic panel, ignore all the pretty colored electronic components. You can't do anything with them anyway. Pay attention only to the wiring terminals (screws) and their labels.

A zoning panel has two purposes. First it gets your wiring organized and keeps it that way. Second it provides the option for one zone, such as domestic hot water, to have priority over all the others.

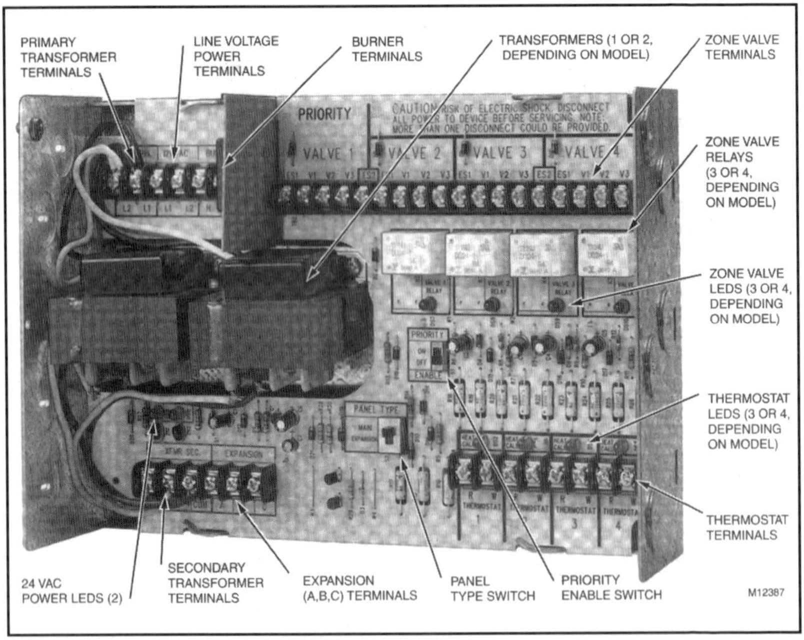

Courtesy of Honeywell

Figure 38. Electronic zoning panel.

Here's how the panel organizes you. You don't have to think about circuits. Just bring in wires like the panel tells you to. Bring line voltage in to the "line voltage power terminals." Wire the zone valves into the "valve" terminals. Wire the thermostats into the "thermostat" terminals. Soon you're done. To make it even better, you don't have to unscramble wires when you troubleshoot.

Let's look at the external wiring diagram. The panel is rotated one quarter turn (90 degrees) from the photograph above. That's probably just because it fits on the page better that way.

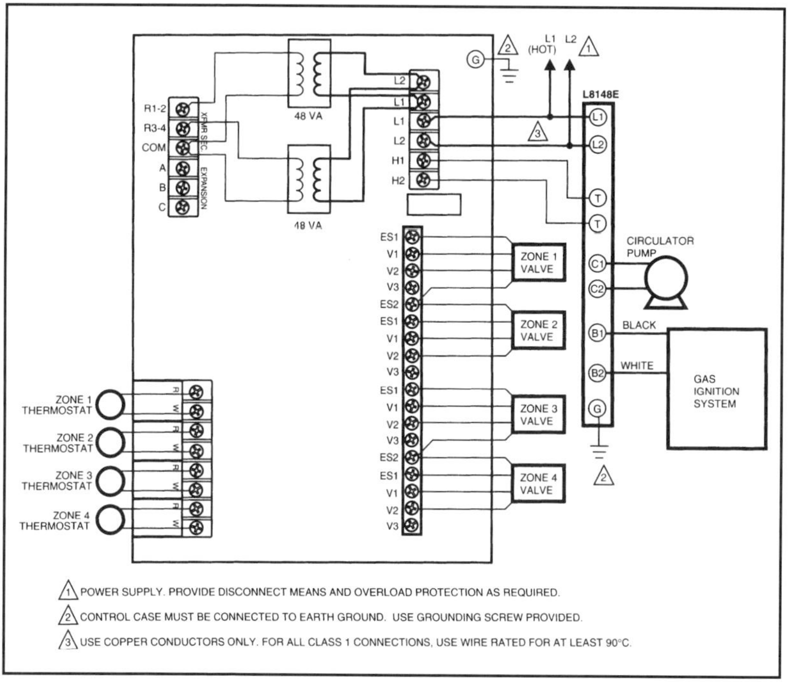

Figure 39. Zoning panel external schematic.

The thermostat terminals are on the left side of the drawing. The zone valve terminals are on the right.

Line voltage comes into L1 and L2 on both the zoning panel and the aquastat relay. There is another set of L1 and L2 terminals on the same terminal strip in the zoning panel. Looking at the control shows that the manufacturer has already made these connections.

Terminals H1 and H2 are shown wired to the T-T terminals of the aquastat relay. You might refer back to the aquastat relay section a few pages back to see some advantages of this configuration.

The zone valves shown are 4-wire valves. When a zone valve has four wires, only two of them are for the motor. Those two connect to V1 and V2. It doesn't matter which motor wire goes to which terminal. Terminals ES1 and ES2 are for the end switch wires.

Check the technical literature for drawings for 2-wire, 3-wire and 5-wire zone valves.

An internal schematic generally isn't provided for an electronic panel.

Zoning panels are available for various numbers of zones, depending upon the model and manufacturer. You can choose one large panel, or several smaller panels. An advantage of several panels is that if one fails, the others can still provide heat to some of the building. Smaller panels can often be wired ("daisy chained") together so that a single zone has priority over all the others. Zoning panels are also available for use with zoning pumps.

An outdoor reset control, also known as an outdoor temperature compensator, is another electronic panel. (For a review of outdoor reset, look at the "Hydronic Controls that Are Switches" chapter.) As with the other panels we have looked at, this panel helps to organize your work. The electronics also makes some options available, such as domestic hot water priority, and warm weather shut-down.

The external wiring diagram shows hook-up to external devices.

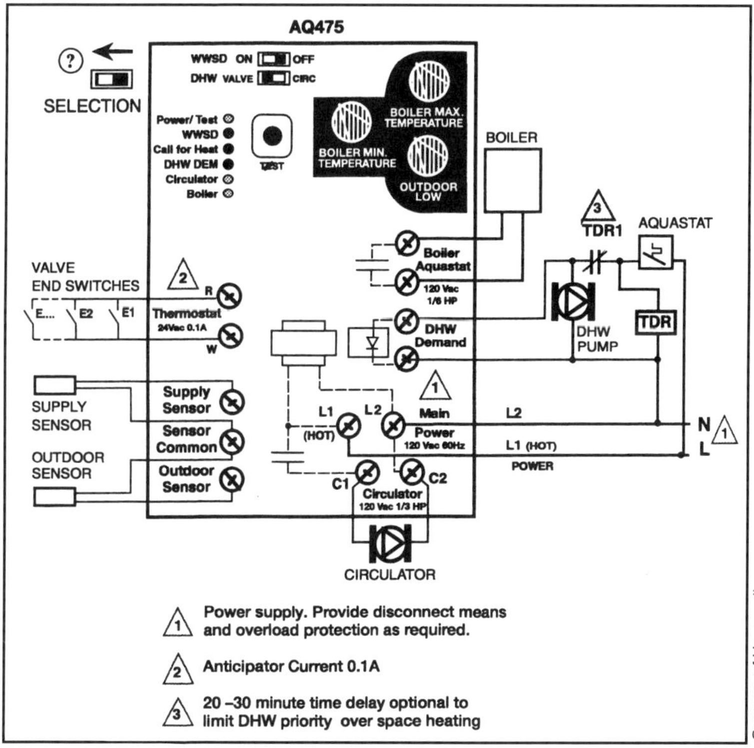

Courtesy of Honeywell

Figure 41. External connection diagram of an electronic outdoor reset controller.

There are just six devices shown—the thermostat, supply sensor, outdoor sensor, circulator, L1-L2 line voltage, and the boiler.

The only thing left after wiring in the external connections is to set the three temperature dials, and to position the two option selection switches at the top of the panel.

Let's look at another electronic outdoor reset control panel.

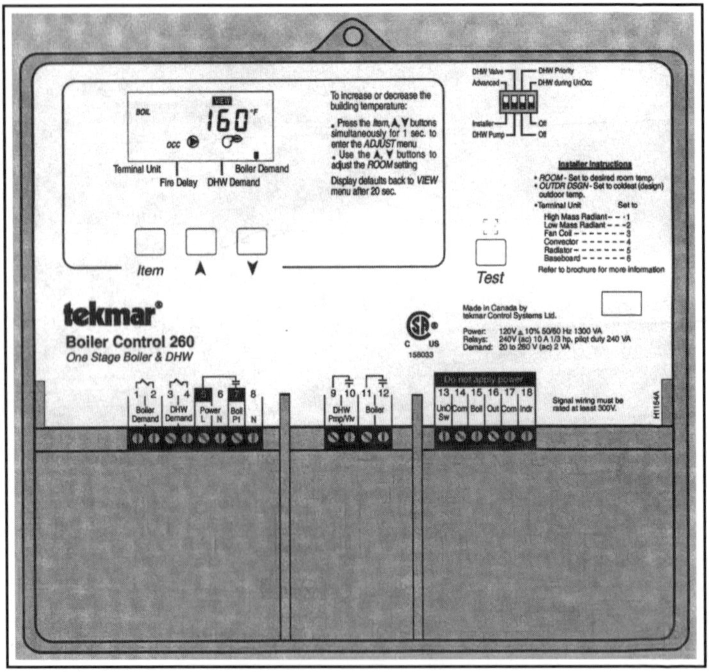

courtesy of tekmar

Figure 42. Another electronic outdoor reset control panel.

Don't forget what you've already learned—a panel's purpose is to make life easier for you, not to confuse you.

Panels start looking more complicated when the terminals are smaller and closer together. Regardless, it's still up to you to think about one thing at a time. Pick a place to start and work around the panel in a systematic manner.

Here's the external connection diagram for this panel.

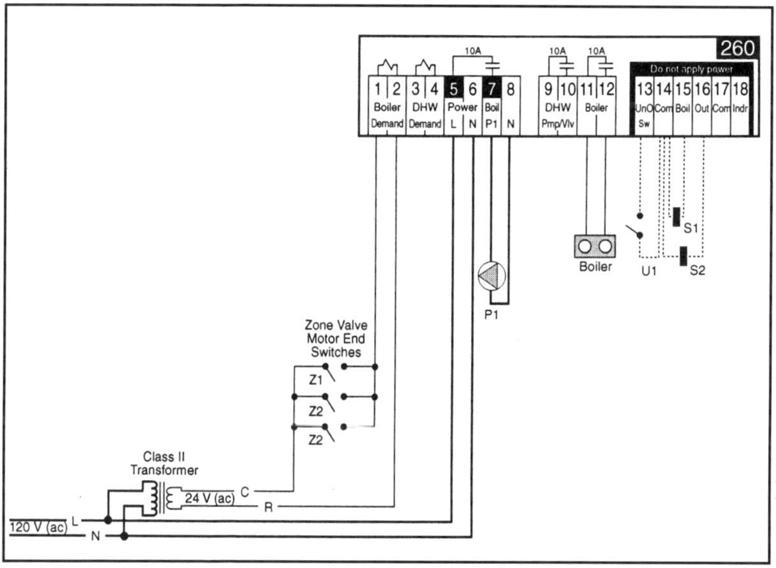

courtesy of tekmar

Figure 43. External connection diagram an electronic outdoor reset controller.

As with the other panels we've looked at, the rectangle outlines the control itself. The lines outside the rectangle show how to wire equipment and devices to it.

Let's start with the power supply. At the bottom left of

the drawing is 120V coming in. It goes two places: to the primary side of the transformer, and to two "power" terminals on the panel.

All we need to do is wire in equipment and devices where we're shown and trust the panel to do what it's designed to do. But let's imagine how the circuits work.

The transformer steps down voltage to 24V. You can see a circuit using this 24V power supply. The switch in the circuit is actually three switches in parallel-labeled "zone valve motor end switches." Because they are in parallel, any one of the three can complete the circuit. The load is across terminals 1 and 2, labeled "Boiler Demand." You can see a relay coil (load) drawn across terminals 1 and 2.

Moving from left to right across the drawing, and across the panel, we see the terminals 3 and 4 are labeled "DHW Demand." If domestic hot water were part of the system, its switching would connect to these two terminals.

Terminals 5 and 6 are connected to 120V power.

Terminals 7 and 8 are connected to the boiler pump. This information is clarified in the notes in the upper left corner of the drawing—"P1 = Boiler System Pump." In this circuit the pump is the load. Power is coming through panel terminals 5 and 6. The switch is drawn just above terminal 7. Since we have a relay coil load drawn above terminals 1 and 2, we can guess that this switch is associated with the relay coil. If this is a relay switch, it will close when the relay coil is powered. The coil will be powered when a zone valve end switch closes. The logic is that when a zone valve end switch closes, the boiler pump comes on.

Terminals 9 and 10 function just like terminals 7 and 8, but for the domestic hot water, if it's included in this system.

Terminals 11 and 12 power the boiler. The switch drawing above the terminals suggests how this works.

Terminals 13 and 14 are connected to an external circuit showing an open switch and "U1." The notes in the upper left corner say that U1 is an optional timer.

Terminal 14 is labeled "com" for "common." It's not surprising to see that three different devices could be wired into terminal 14. Besides the timer to terminal 13, terminal 14 could serve as the common for the boiler sensor ("S1") to terminal 15. It could also serve for the outdoor sensor ("S2") to terminal 16.

Terminals 17 and 18 aren't used, so we're not going to be concerned about them.

One of the greatest fears we have about panels is "blowing them up." Put power only where the diagram tells you to, and nowhere else. Apply only the voltage that the diagram tells you to, and no more. Hey, it's like so many other things—if you follow the instructions, think about what you're doing, and take your time, it'll probably work!

Conclusion

Now you have seen many hydronic controls. Isn't it great to know that there aren't really all that many different kinds! They all come down to three possibilities—power supply, switch, or load. Even when they're combined together, it's still the same three.

When you look at a circuit, think of a complete circle. If it looks more like a pile of spaghetti, think of a bug on a rope. "If I were a bug, could I go in one direction along this rope and get back to where I started?"

And when you see a panel, you'll have a good idea what's inside that box before you even open it up. It's power supply, switches and loads ready to be wired with other switches and loads to make circuits.

Use this information well, and share it with a friend.

The end.

Index

Order Information

To order copies of

Quick & Basic Electricity
Quick & Basic Hydronic Controls
Quick & Basic Troubleshooting

Mail this form to:
P.I.G. Press
759 E. Phillips Drive S.
Littleton, CO 80122-2873

Enclose $20.00 U.S. funds for each book plus $3.95 for shipping and handling. (Price is subject to change.)

Indicate which book(s) you wish to order and the quantity of each.

Your Name: ______________________________

Your Address (include zip code): ________________

____

____

Phone (if we have a question): _________________

Please also contact us for video tapes or information about seminars by the author at your location

Questions? Call P.I.G. Press at:
303-795-2679
or Fax 303-795-9350

Thank you